ENERGY
IT'S PAST, IT'S PRESENT, IT'S FUTURE

By Martin J. Gutnik

Illustrated by Sam Shiromani

CHILDRENS PRESS, CHICAGO

To Laurie

ACKNOWLEDGEMENTS

Margaret Berg—Teacher and typist, Shorewood, Wisconsin
Laurie Gutnik—My dear wife and mother of my children
Donald Neuman, Ph.D.—Consultant and advisor

Library of Congress Cataloging in Publication Data

Gutnik, Martin J
   Energy: its past, its present, its future.

   SUMMARY: Defines various types of energy and its uses.
   1. Power resources—Juvenile literature.
[1. Power resources] I. Shiromani, Sam. II. Title.
TJ16312.G87   333.7   75-11825
ISBN 0-516-00525-1

Copyright © 1975 by Regensteiner Publishing Enterprises, Inc.
All rights reserved. Published simultaneously in Canada.
Printed in the United States of America.

1 2 3 4 5 6 7 8 9 10 11 12 R 78 77 76 75

WHAT IS ENERGY?

Have you ever been told to eat your breakfast because it will give you energy? Have you ever felt tired and said you didn't have enough energy to play?

Energy is the ability to do work. If we did not have energy, work would not be done. Energy makes things move. Energy makes things grow. Energy is responsible for every action that takes place.

ENERGY FROM THE SUN

Most energy on earth comes from the sun. This is energy from the sun's rays. It is called RADIANT ENERGY. The sun's rays give light and heat to the earth. They warm the air and cause winds to blow.

Fusion

The sun receives its energy from explosions that take place in the center of the sun. These explosions are caused by particles inside the sun coming together. Particles are very small pieces of material, some so small that they cannot be seen.

When particles come together it is called FUSION. To fuse means to come together with a large amount of heat. When particles in the sun fuse, they create a large amount of heat and light.

The heat and light created in the sun travels down to the earth in *rays*, or *waves*. This is why energy from the sun is called radiant energy.

The radiant energy from the sun gives us food, wood, coal, oil, natural gas, wind, and flowing water. We use all of these things for energy on earth.

Green Plants

Green plants use the sun's energy to make food. Some animals eat the green plants for their food. Other animals eat animals that eat green plants.

If there were no energy from the sun, then green plants could not make food. If green plants could not make food, most life on earth could not exist.

Trees

Trees use the sun's energy to make food and grow. As they grow, trees produce wood. When we start a fire, the energy of the sun stored in the wood gives us heat.

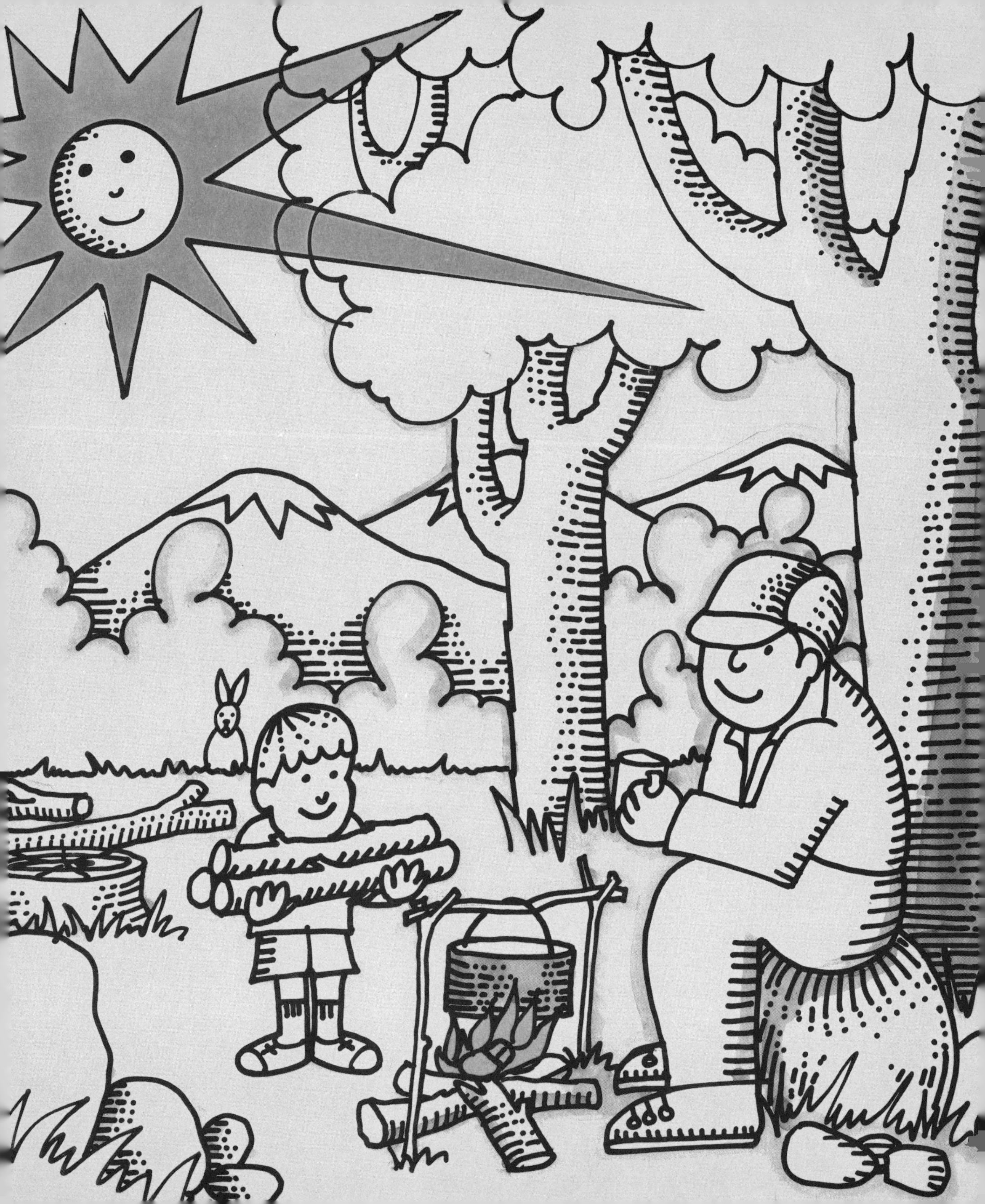

SOIL

ROCK

PEAT

COAL

Coal

The radiant energy of the sun gives us coal. Billions of years ago, plants and trees grew in thick forests and swamps. When these plants died, some of them fell into the swamps and were covered with water. They decayed and became *peat*. (Peat is the first material in the making of coal.) This peat was pressed down by other plants and by the earth. Over the years it was pressed into layers. Each layer had a different hardness. These layers of pressed-down peat formed a product called coal.

When we burn coal we are releasing the energy that had been stored in these plants. These plants stored the energy of the sun. When coal is burned, the sun's energy is being released to do work for us.

Oil

    Hundreds of millions of years ago, many small plants and animals lived in shallow seas and marshes. When these plants and animals died, their bodies were decomposed —broken down. As the years went by they were pressed down under layers of rock. Today these plants and animals give us oil.

    As we burn products made from oil for heating homes, running automobiles, or running power plants, we are releasing energy from the sun. This energy was made and stored millions of years ago and is being used today to do work for us.

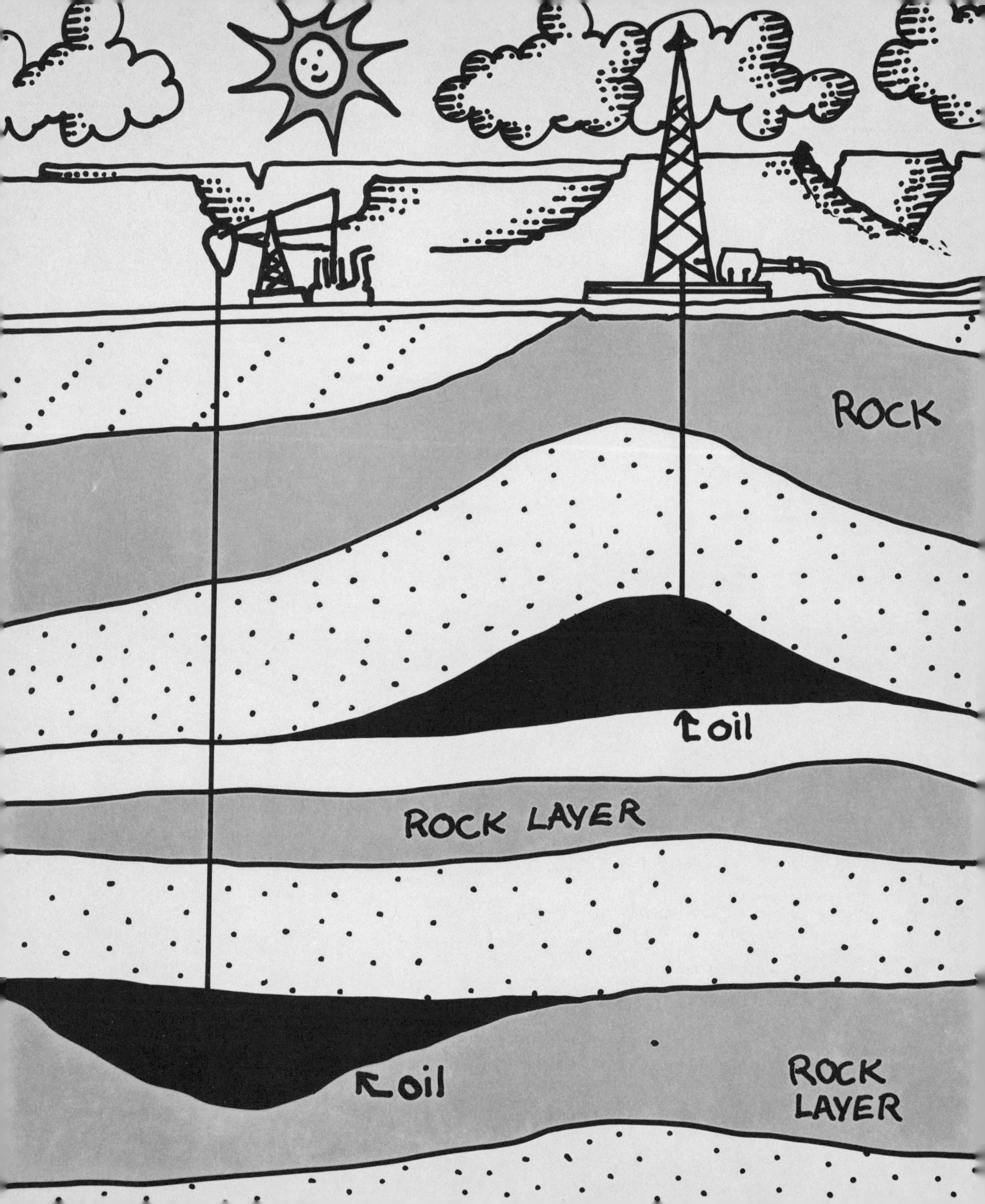
ROCK
oil
ROCK LAYER
oil
ROCK
LAYER

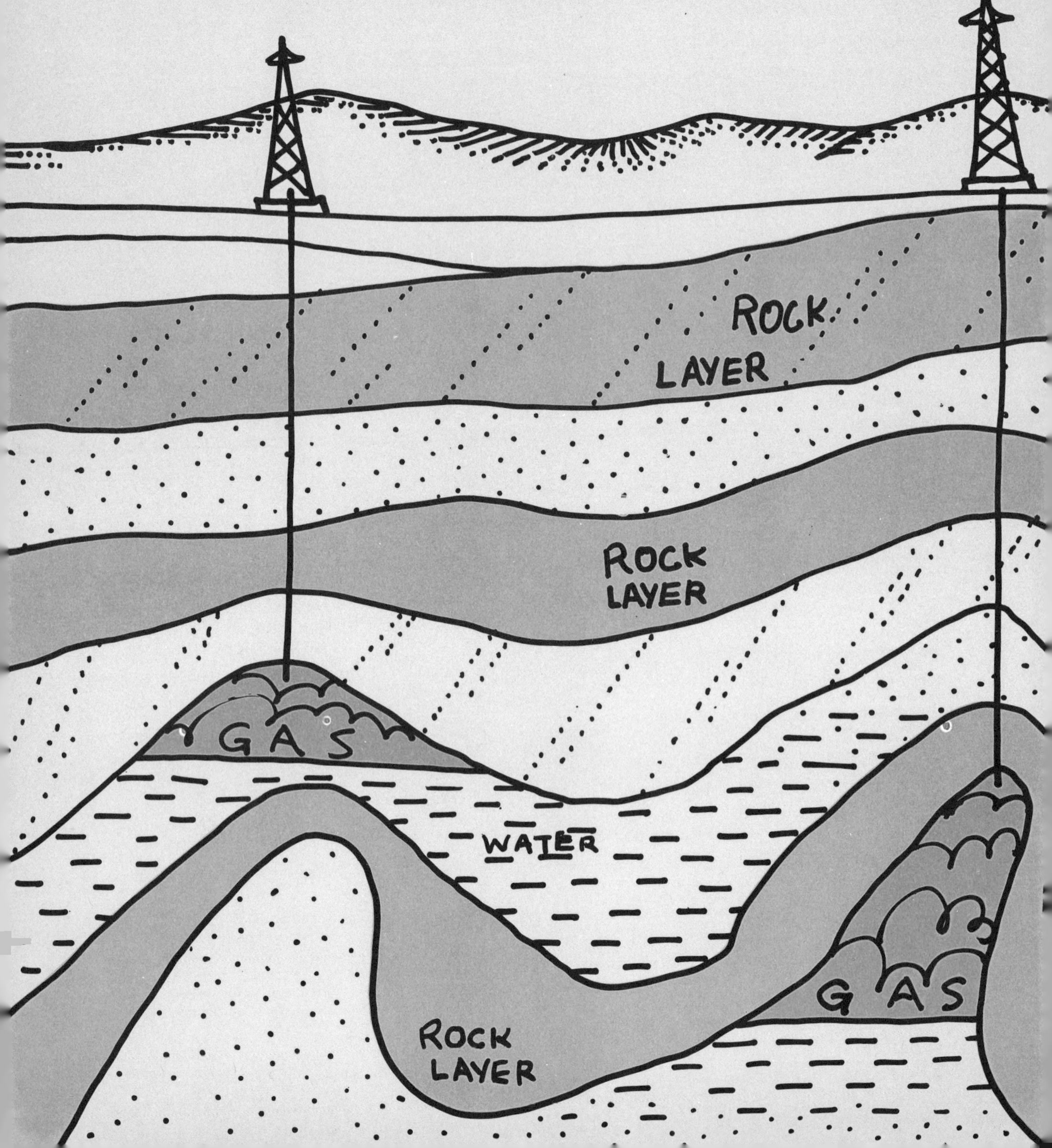
ROCK
LAYER
ROCK
LAYER
GAS
WATER
GAS
ROCK
LAYER

Natural Gas

Natural gas was formed in much the same way that oil was formed. It is often found with or near oil.

The animals and plants that formed natural gas used energy from the sun for food. When they died and were covered over, this energy was stored with them. When we burn natural gas today we are using energy from the sun. This energy was created millions of years ago.

Flowing Water

As the sun's rays shine down on lakes and oceans they cause water to evaporate (go into the air). The water in the air forms clouds. The clouds allow rain to fall. When the rain falls it causes rivers to flow. We use flowing rivers and their waterfalls as natural energy for power plants. We also build dams on rivers. Dams hold back the rivers' water. When the dam is opened up it creates a waterfall which is used as energy to run power plants. This energy comes from the sun.

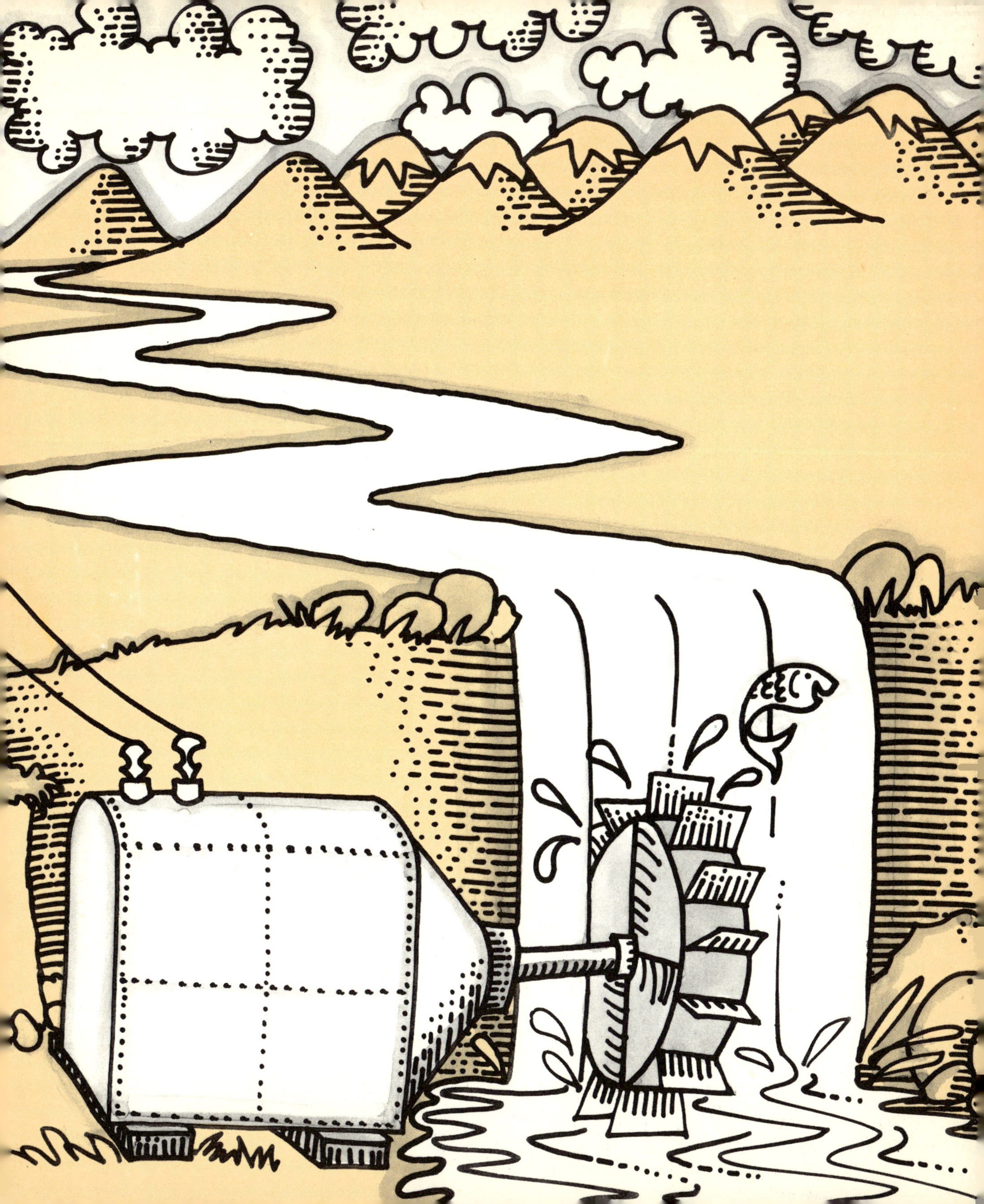

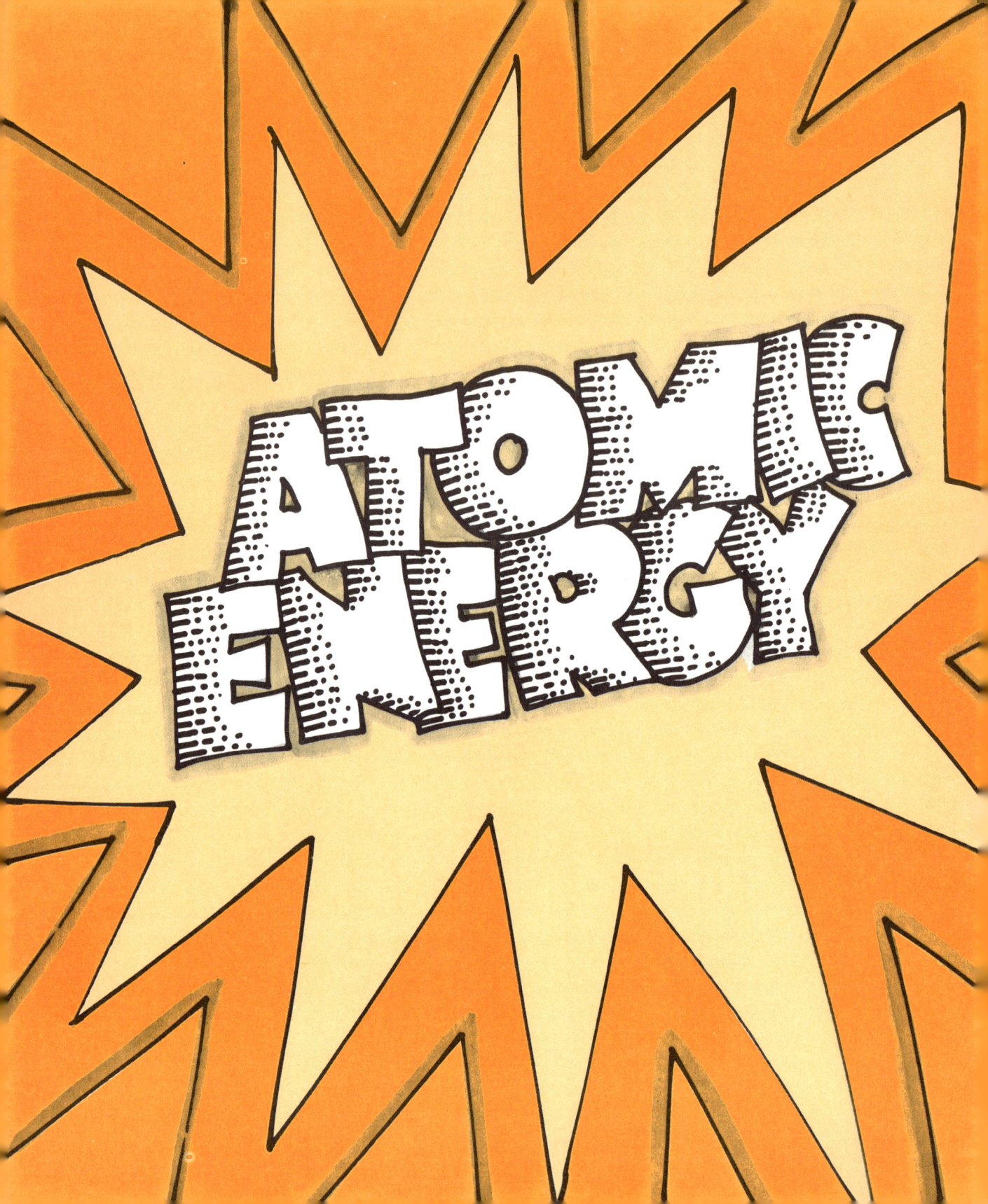

ATOMIC ENERGY

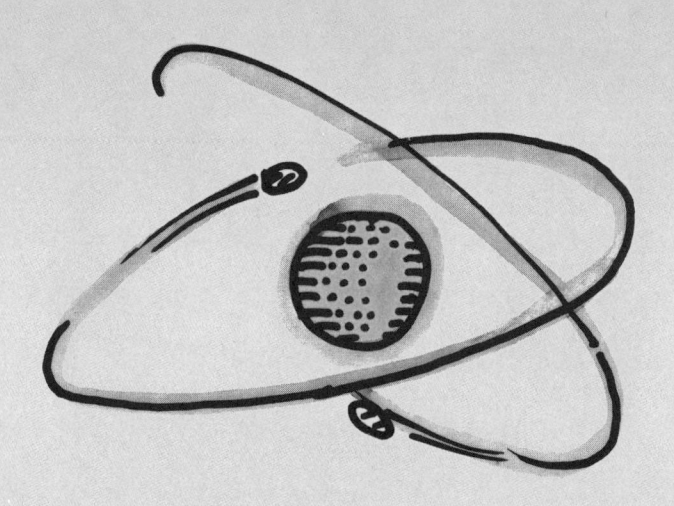

ATOMIC ENERGY

For many years people believed that all of their energy came from the sun. In the early part of this century, scientists found out that there is also energy on earth.

This energy is ATOMIC ENERGY. Scientists discovered that certain elements split apart very easily. When these elements split, they create a great deal of heat energy. This is called FISSION. Fission is the splitting of an element with the release of a great deal of heat.

Atomic energy is now being used to operate nuclear submarines, electrical power plants, and many other machines.

As man's knowledge of atomic energy increases so will his use of this energy.

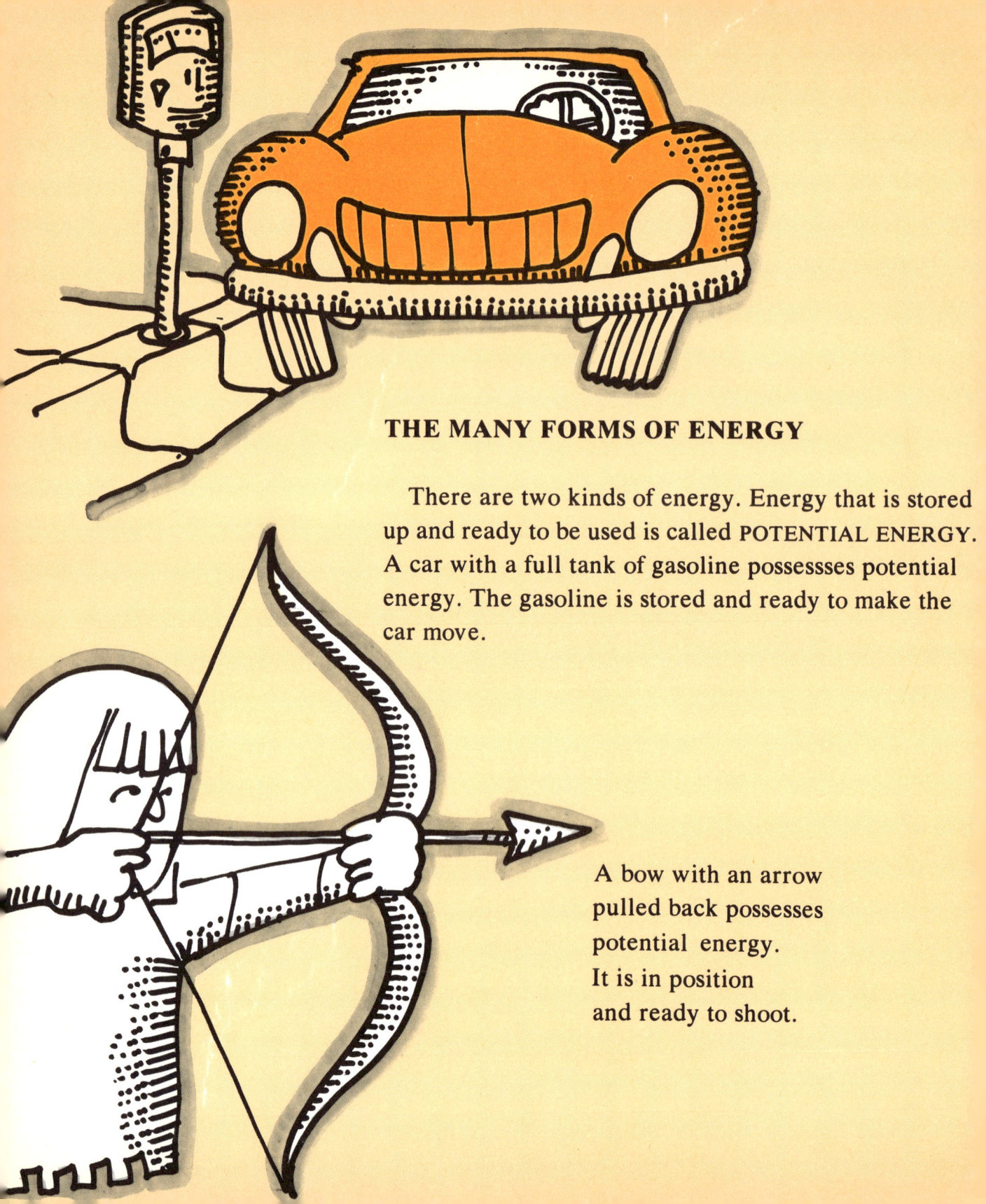

THE MANY FORMS OF ENERGY

There are two kinds of energy. Energy that is stored up and ready to be used is called POTENTIAL ENERGY. A car with a full tank of gasoline possessses potential energy. The gasoline is stored and ready to make the car move.

A bow with an arrow pulled back possesses potential energy. It is in position and ready to shoot.

Energy that is moving is called KINETIC ENERGY. A car moving down the road possesses kinetic energy.

An arrow heading for its mark possesses kinetic energy.

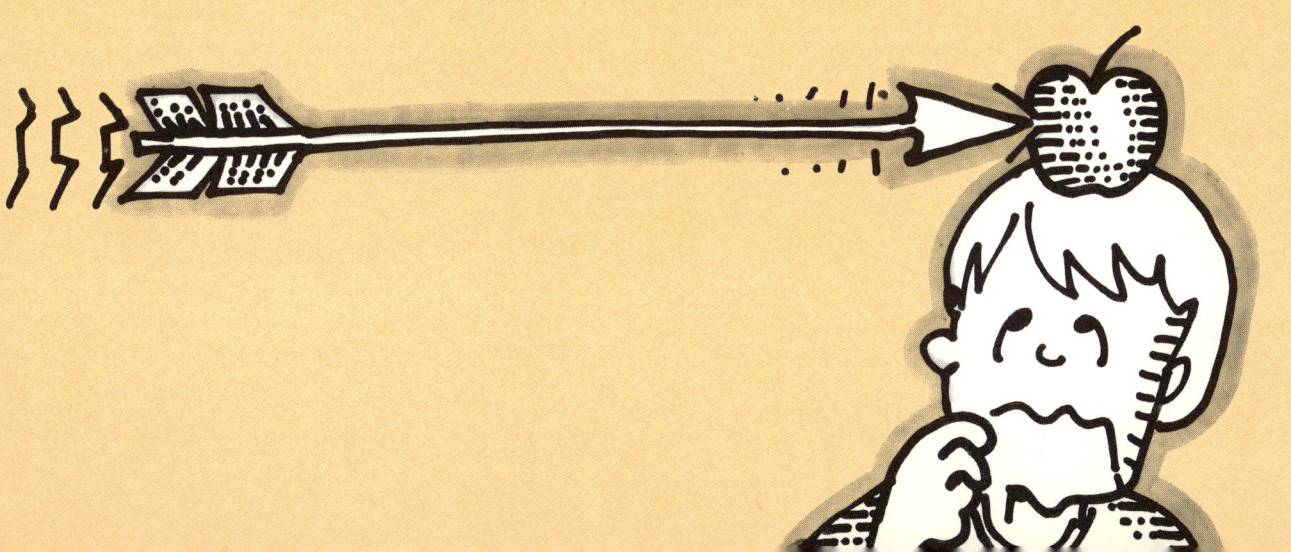

ATOMIC ENERGY

CHEMICAL ELECTRICAL ENERGY

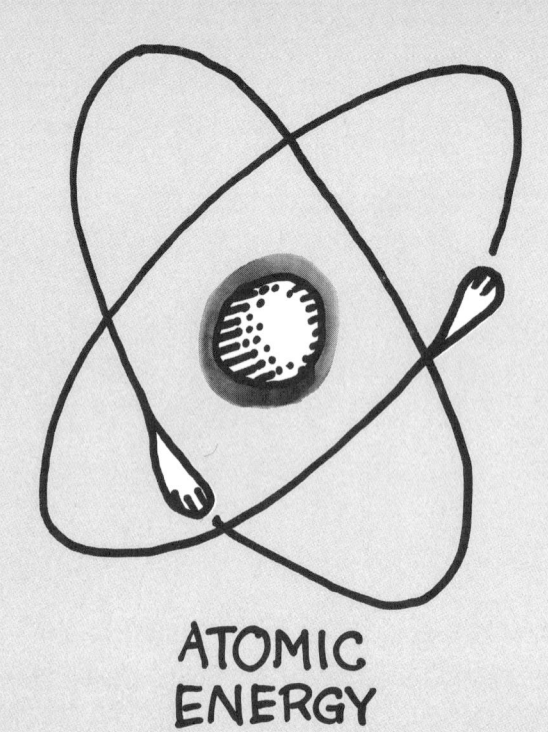

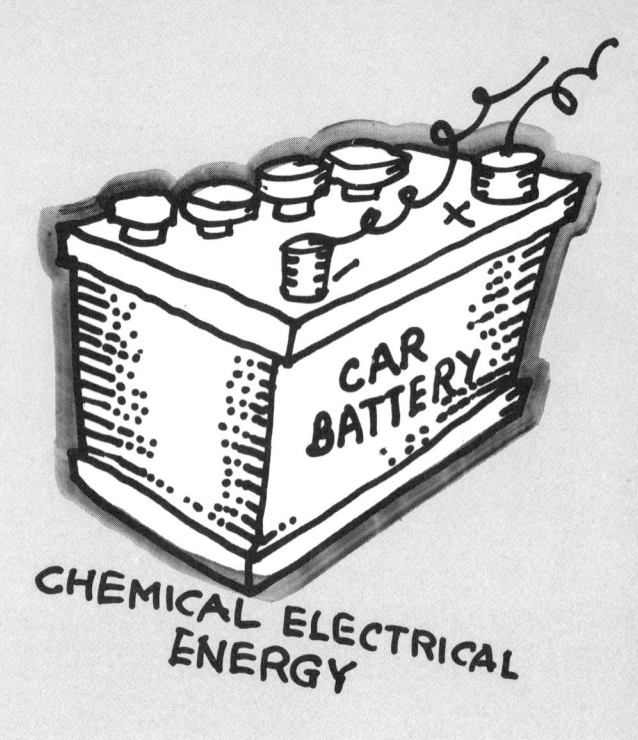

Energy takes many different forms.

Energy can be changed from one form to another, but it cannot be created or destroyed.

Energy comes in many different forms and can be changed from one type to another.

ELECTRICAL HEAT & LIGHT ENERGY

CHEMICAL ENERGY (BURNING)

MUSCLE AND MECHANICAL ENERGY

CHEMICAL ENERGY (FOOD MAKING)

RADIANT ENERGY

WATER ← STEAM →

HEAT

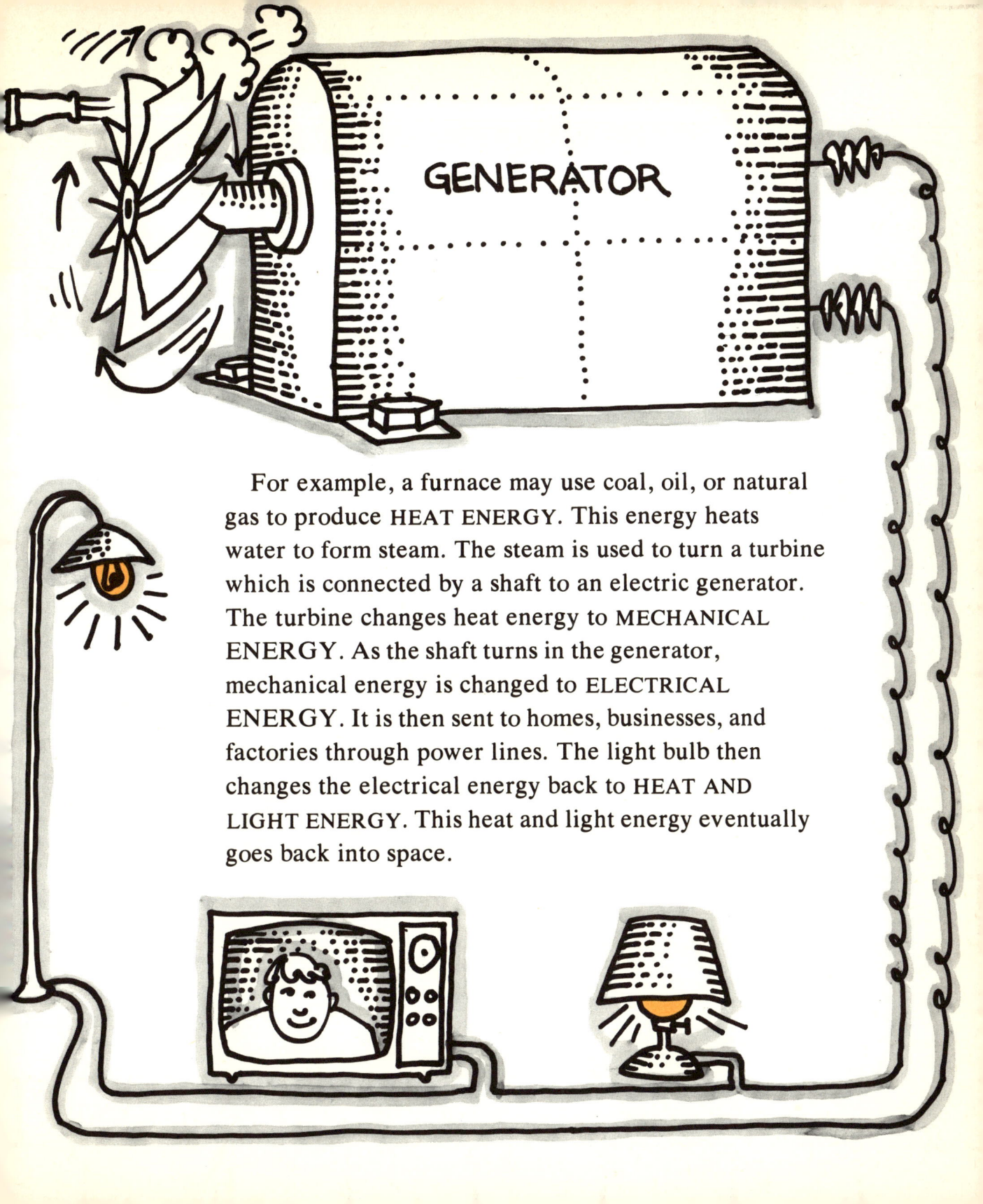

For example, a furnace may use coal, oil, or natural gas to produce HEAT ENERGY. This energy heats water to form steam. The steam is used to turn a turbine which is connected by a shaft to an electric generator. The turbine changes heat energy to MECHANICAL ENERGY. As the shaft turns in the generator, mechanical energy is changed to ELECTRICAL ENERGY. It is then sent to homes, businesses, and factories through power lines. The light bulb then changes the electrical energy back to HEAT AND LIGHT ENERGY. This heat and light energy eventually goes back into space.

HOW PEOPLE HAVE DEVELOPED AND USED ENERGY

Man's standard of living has gone up as he has found new energy sources. As man has progressed, his need for more and better forms of energy has increased.

The first energy that people used for their work was MUSCLE ENERGY. The only work they could do was work their muscles could do for them.

Later people discovered how to control fire. They used fire for warmth and protection from wild animals. They also found out that meat tasted better when cooked in a fire.

Crude tools were developed to help people with some of their work. This was the first form of MECHANICAL ENERGY.

Their tools were mainly rocks and sticks of certain shapes and sizes.

As the years passed these tools were improved. They were shaped into the forms people wanted and handles were put on them to make them easier to use.

Human population began to increase. People became civilized. They began living in small towns or in small farming communities.

It was discovered that plants grow better when the soil is tilled (broken up). Machines were developed for tilling the soil. The first machines were pushed by muscle energy. People then used animals for energy to till the soil. Finally machines using CHEMICAL ENERGY (in the form of gasoline) were developed to do this work.

The discovery of coal, oil, and natural gas and the invention of machines that used these fuels started the rapid growth of industry, transportation, and cities.

With this rapid growth came the cry for more machines and more energy to run them.

ENERGY PICTURE TODAY

Automobiles, jet planes, home and industrial heating and power, air conditioners, and the millions of other electrical gadgets created by man have greatly increased the world's use of and demand for energy.

Coal, oil, and natural gas are the main sources of fuel used to supply this energy. You remember that coal, oil, and natural gas were formed from plants and animals that died billions of years ago. These plants and animals are called FOSSILS. Thus, coal, oil, and natural gas are called FOSSIL FUELS. Today over 90 percent of the world's energy comes from fossil fuels.

FOSSIL FUEL 90%

coal

oil

natural gas

OTHER FUEL SOURCES: NUCLEAR, SOLAR, ETC.

HOW MUCH FUEL IS LEFT?

Fossil fuels were formed hundreds of millions of years ago. It took millions of years for them to form. They are still forming today, but very slowly.

We are using up our fossil fuels much faster than they are being formed. This means that when present-day reserves are used up they will be gone for many years to come.

At the present rate of use there is about a five-hundred-year supply of coal left.

At the present rate of use there is about a thirty-year supply of oil left.

At the present rate of use there is about a twenty-year supply of gas left.

WORLD SUPPLY OF FOSSIL FUEL

Some countries will run out of fossil fuel energy sooner than others. This is beginning to happen already. The major industrialized countries use more fuel than other countries do. Because they do not have all the sources of this fuel within their borders thay have to import some of their fuel. They become dependent on other countries for their energy.

USA + CANADA 20% ★

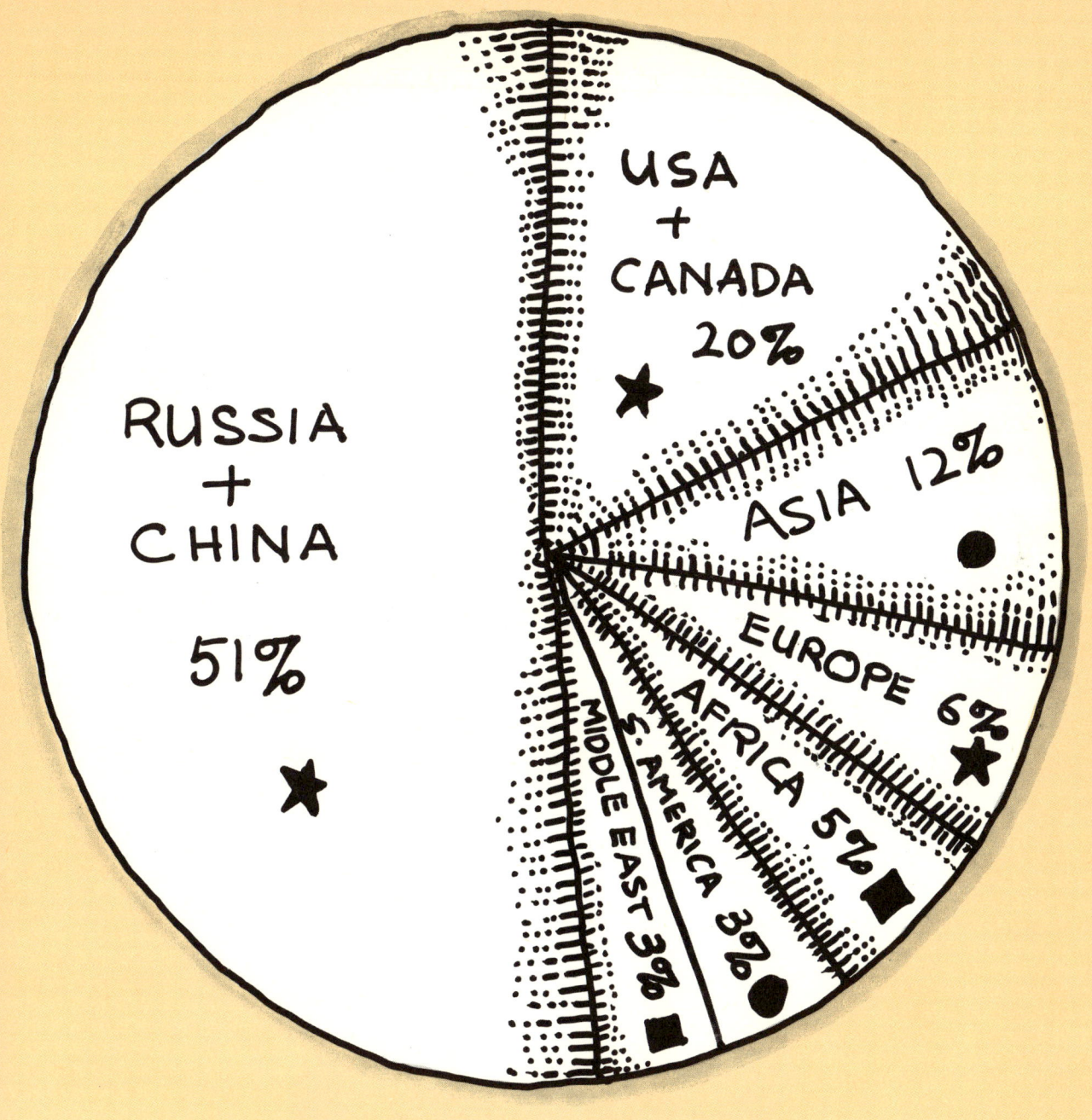

RUSSIA + CHINA 51% ★

ASIA 12% ●

EUROPE 6% ★

AFRICA 5% ■

S. AMERICA 3% ●

MIDDLE EAST 3% ■

★ HIGH INDUSTRY
● DEVELOPING
■ VERY LITTLE INDUSTRY
% AMOUNT OF FUEL RESERVE IN AREA

ENERGY PICTURE FOR THE FUTURE

Most of the fossil fuel we use for energy today will be gone by the year 2000. What other sources of energy will be available to us then?

Scientists today are beginning to work on energy sources that will not run out and will not pollute. The world is beginning to run out of energy and they know they must find an answer soon.

NUCLEAR REACTOR

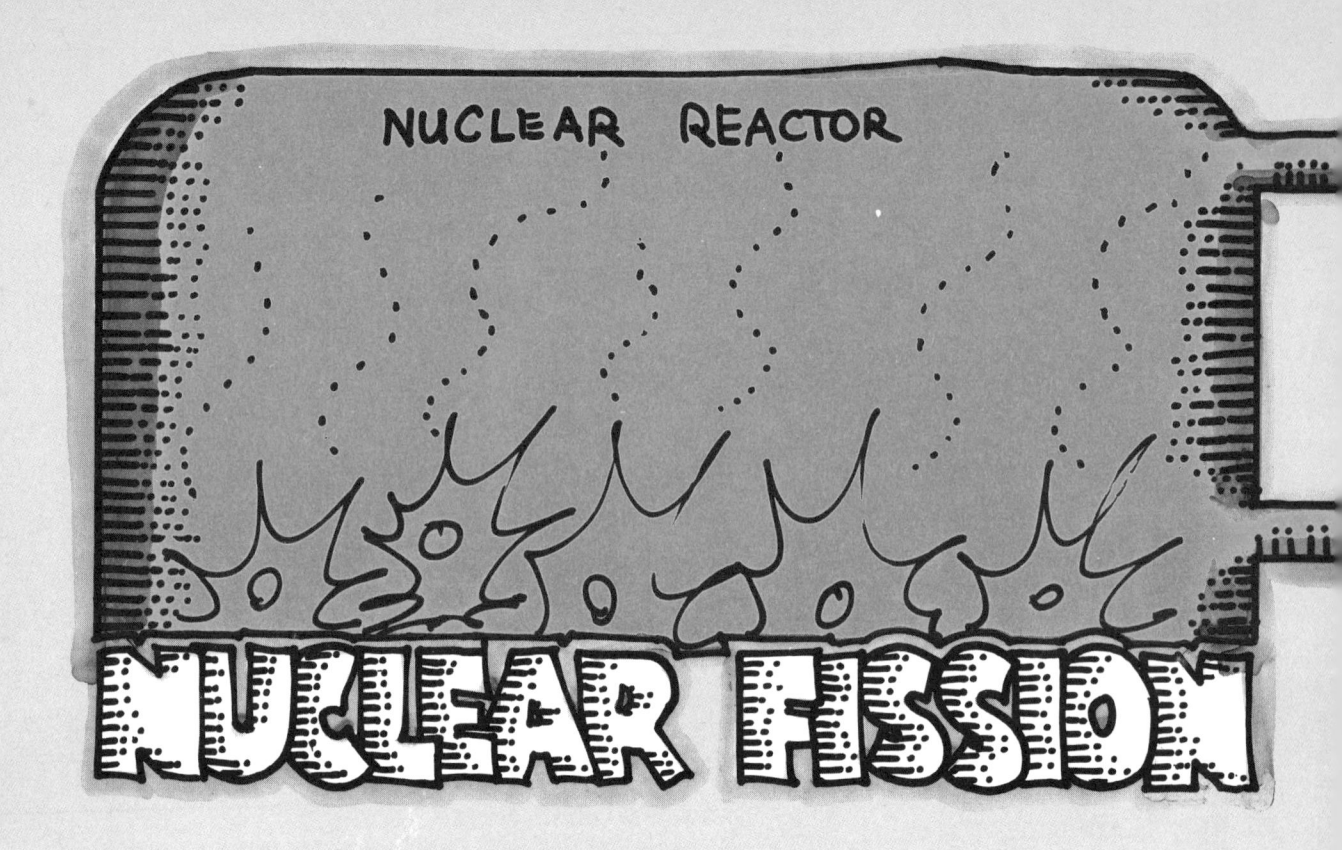

NUCLEAR FISSION

One answer for our future needs could be greater use of ATOMIC ENERGY. You remember that atomic energy is formed when elements split to produce heat. Scientists have learned to control the splitting of these elements. They use an atomic reactor. This reactor acts in the same way a coal furnace does. The elements split inside the reactor and produce heat. The heat is used to turn water into steam. The steam operates an electrical generator to give us electricity.

Scientists are now working on an atomic reactor that will make its own fuel at the same time it is producing heat energy. This is called an atomic breeder reactor. When they perfect this reactor it could solve many of our energy needs.

HEAT EXCHANGER

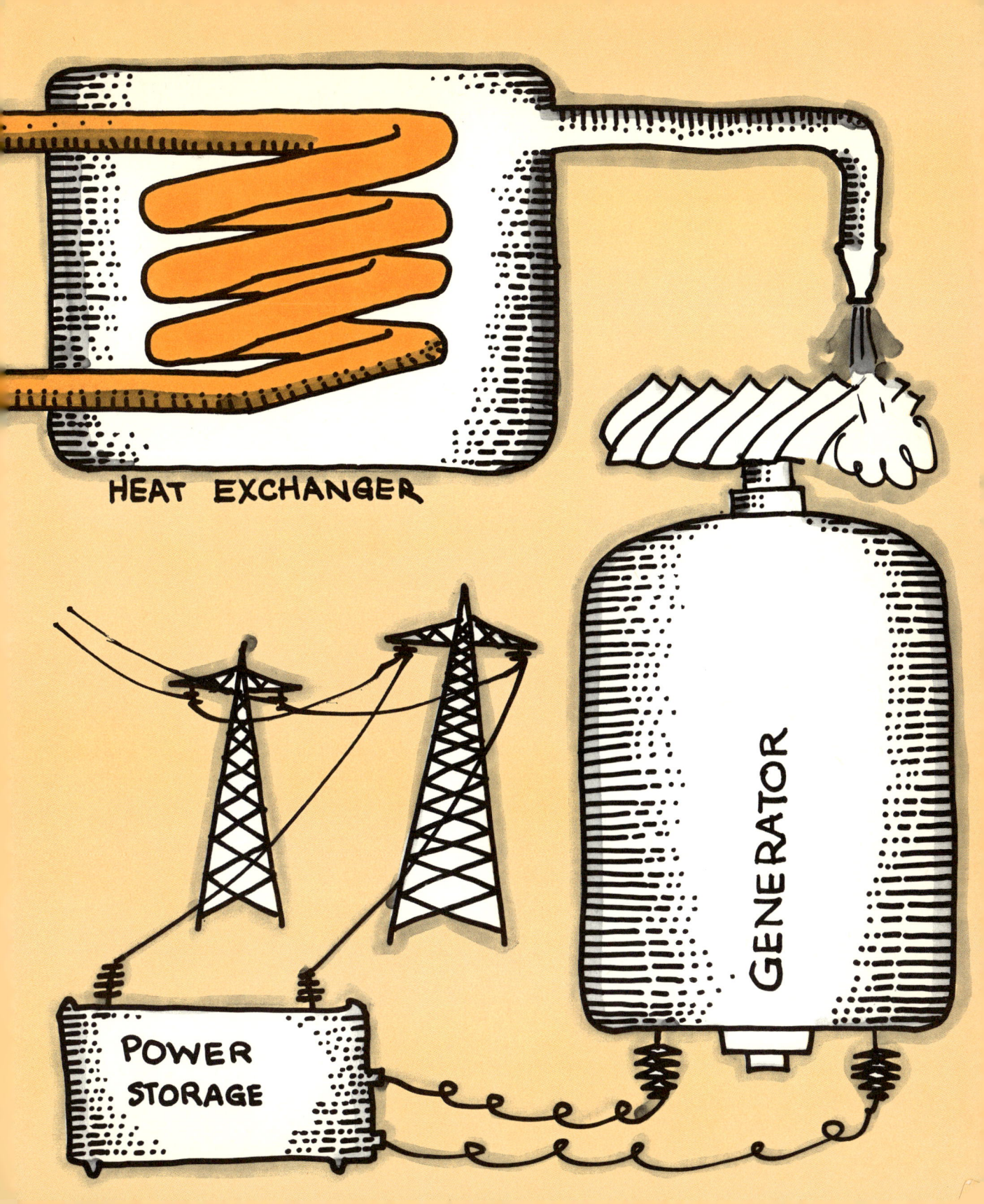

GENERATOR

POWER STORAGE

GEYSER

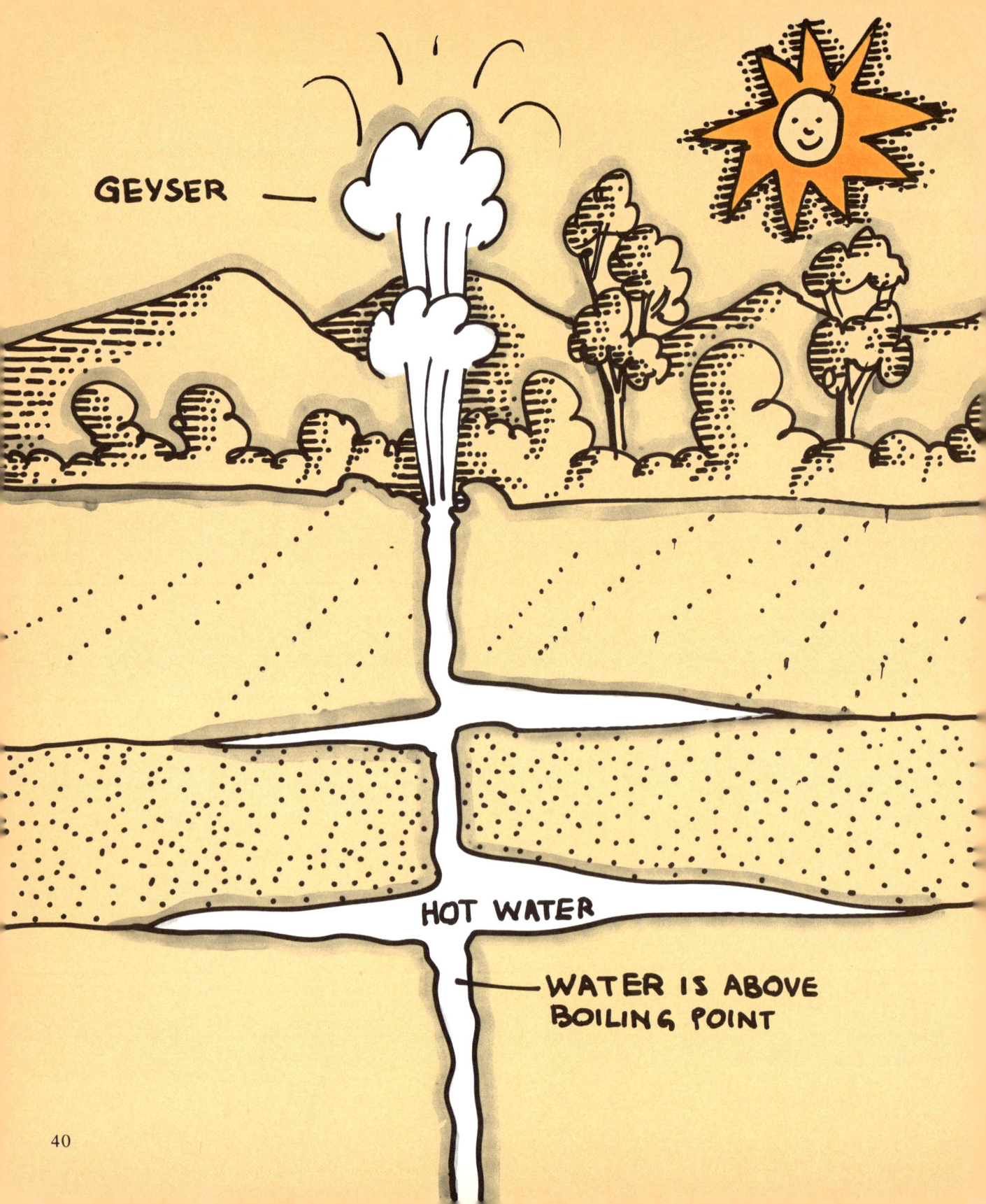

HOT WATER

WATER IS ABOVE BOILING POINT

GEOTHERMAL ENERGY

Another form of energy that could be harnessed for future use is GEOTHERMAL ENERGY. This energy comes from the natural heat of the earth. For example, water seeps through cracks in the earth's surface and works its way deep into the earth. Here, because of great natural heat and pressure, the water becomes very hot and expands. When this hot water finds its way back up to the earth's surface, it shoots from the earth with explosive force. We call these columns of steam geysers. The heat given off by geysers and water spouts could be used to operate electrical generators and provide electricity for major areas of the earth.

SOLAR ENERGY

SOLAR ENERGY (energy from the sun) might be part of the solution for future energy sources.

As much energy falls on the earth from the sun in one minute as is used in a year. But this energy is spread very thin and is difficult to capture and concentrate.

Scientists are developing solar collectors. These would collect the energy rays from the sun and use them to heat water to warm homes.

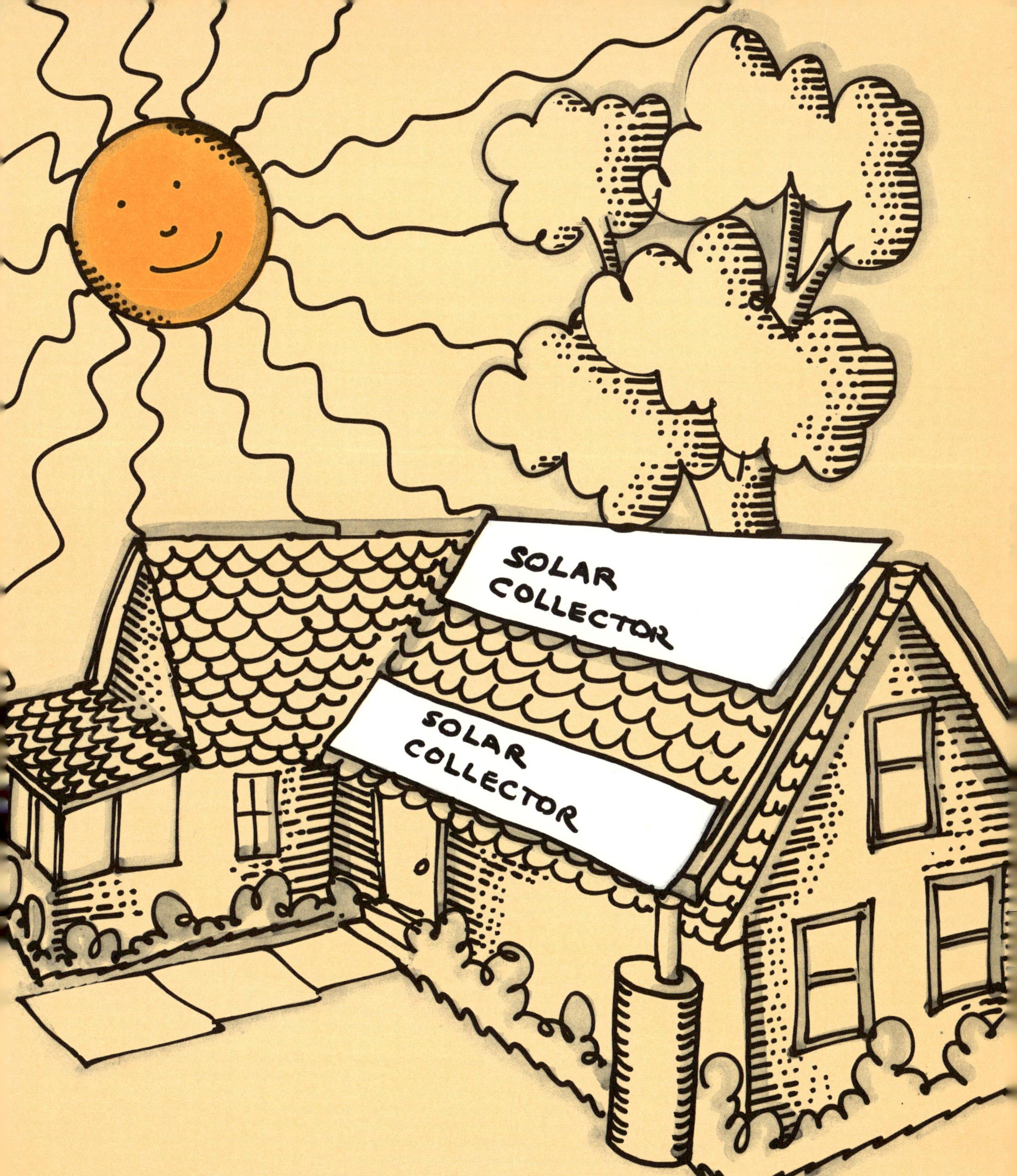
SOLAR COLLECTOR
SOLAR COLLECTOR

FUSION

You remember that the sun receives its energy from FUSION.

Scientists today are working on ways of creating a fusion reaction on earth. This type of reaction requires very high temperatures in order to occur.

If scientists can accomplish a fusion reaction and control it we could have a new major source of energy. It would be like having the sun right here on earth.

As you have seen, energy has been a major factor in shaping people's way of living ever since prehistoric times. How people live in the future will be determined by the discovery and use of new energy sources.

GLOSSSARY OF TERMS

1. *atomic breeder reactor*: A reactor that makes more fuel while producing energy.
2. *atomic reactor*: A machine used to control the splitting of elements to make energy.
3. *decompose* (dee cum POZE): To break down into smaller particles.
4. *energy* (EN uhr jee): The ability to do work.
5. *fission* (FISH un): The splitting of elements.
6. *fossil fuel* (FAHS uhl): Any fuel that is formed from animal or plant remains.
7. *fusion* (FYU zhun): When particles come together with a large amount of heat.
8. *kinetic energy* (keh NET ik): The energy of movement.
9. *potential energy* (puh TEN chul): Energy that is stored and in position to be used. The energy of position.
10. *radiant energy* (RAID ee ent): Energy that travels in waves. Light and heat from the sun.
11. *solar energy* (SO luhr): The radiant energy from the sun.

About the Author

Martin Gutnik, an innovative elementary school science teacher, lives with his wife and two young children in Milwaukee, Wisconsin. *Energy, Its Past, Its Present, Its Future,* is his fourth book for Childrens Press. His first three books, *First Experiments in Science and Nature,* are based on a series of experiments on ecology and pollution that he created for his students to perform in the classroom. Though Mr. Gutnik spends much of his time helping his students learn to build ecosystems, develop film, and dissect frogs, his life is not totally surrounded by science. He also collects records from the fifties and enjoys singing folk songs as he accompanies himself on the guitar, which he taught himself to play.

About the Artist

Born in India, Sam Shiromani has made Chicago his home. A dropout from the world of advertising, he devotes most of his time to free-lancing art and photography.